BUILDING BLOCKS OF GEOGRAPHY

EARTH IN SPACE

Written by Alex Woolf

Illustrated by Emiliano Migliardo

WORLD BOOK

a Scott Fetzer company
Chicago

World Book, Inc.
180 North LaSalle Street
Suite 900
Chicago, Illinois 60601
USA

For information about other World Book publications, visit our website at **www.worldbook.com** or call **1-800-WORLDBK (967-5325)**.
For information about sales to schools and libraries, call 1-800-975-3250 (United States), or 1-800-837-5365 (Canada).

© 2023 World Book, Inc. All rights reserved. This volume may not be reproduced in whole or in part in any form without prior written permission from the publisher.

WORLD BOOK and the GLOBE DEVICE are registered trademarks or trademarks of World Book, Inc.

Library of Congress Cataloging-in-Publication Data for this volume has been applied for.

Building Blocks of Geography
ISBN: 978-0-7166-4275-6 (set, hc.)

Earth in Space
ISBN: 978-0-7166-4278-7 (hc.)

Also available as:
ISBN: 978-0-7166-4288-6 (e-book)

Printed in India by Thomson Press (India) Limited, Uttar Pradesh, India
1st printing June 2022

WORLD BOOK STAFF
Executive Committee
President: Geoff Broderick
Vice President, Editorial: Tom Evans
Vice President, Finance: Donald D. Keller
Vice President, Marketing: Jean Lin
Vice President, International: Eddy Kisman
Vice President, Technology: Jason Dole
Director, Human Resources: Bev Ecker

Editorial
Manager, New Content: Jeff De La Rosa
Associate Manager, New Product: Nicholas Kilzer
Sr. Editor: Shawn Brennan
Proofreader: Nathalie Strassheim

Graphics and Design
Sr. Visual Communications Designer: Melanie Bender
Sr. Web Designer/Digital Media Developer: Matt Carrington
Coordinator, Design Development: Brenda Tropinski

Acknowledgments:
Writer: Alex Woolf
Illustrator: Emiliano Migliardo
Series advisor: Marjorie Frank

Developed with World Book by White-Thomson Publishing LTD
www.wtpub.co.uk

Additional spot art by Shutterstock
36-37: UCO/Lick Observatory

TABLE OF CONTENTS

Planet Earth 4
Earth's Spheres 6
Earth's Place in Space 12
Earth's Orbit 14
Earth's Rotation 16
Earth's Tilt and Latitude 18
Seasons 20
Earth's Moon 26
Eclipses 30
Humans in Space 32
Activity: Explore the Phases of the Moon! 36
Can You Believe It?! 38
Words to Know 40

There is a glossary on page 40. Terms defined in the glossary are in type **that looks like this** on their first appearance.

PLANET EARTH

Hey, I'm **Earth!**

I'm your planet ... but how well do you really know me?

Let me take you on a tour of your home (that's me!) and my home (our **solar system**).

I'm a pretty special planet because I'm the only place in the **universe** that is home to humans (and life in general)!

Living things aren't my only interesting feature. I've also got rock ...

... water ...

... and an atmosphere! Hey Atmosphere!

We'll take a closer look at these four features in a minute.

Catch you later!

The study of Earth's physical features and places and the ways in which humans interact with them is known as geography.

All life on Earth is affected by my location, motions, characteristics, and features. So, come along and take a close look at an amazing planet—me!

5

EARTH'S SPHERES

Let's start with a really important sphere—me!

GRAVITY

Like most other planets, I'm spherical because my **gravity** pulls matter in toward my center.

I'm not a perfect sphere, though. The way in which I spin on my axis makes me slightly flattened at the poles ...

... and gives me a bit of a bulge around the middle! We mark the center of my surface with an imaginary line called the equator.

Earth's **circumference** is just over 24,900 miles (40,000 kilometers) around the equator.

At an average walking pace of 3 miles (4.8 kilometers) per hour, it would take you just over 345 days of nonstop walking to get all the way around!

Remember Earth's features mentioned before? These are divided into four systems called spheres.

Sphere 1: Lithosphere

The lithosphere includes all rock and land on and below my surface.

This one can get quite explosive!

Hey, I'm Land. I'm just one part of Earth's lithosphere.

This uneven, solid outer layer is known as the **crust**.

Under the crust, there are three other layers of rock hidden away: the mantle, the outer core, and the inner core.

- CRUST
- MANTLE
- OUTER CORE
- INNER CORE

The mantle is the thickest layer. The rock here is molten (melted) and feels a bit like caramel.

It doesn't taste like it though!

The outer core and inner core are made out of iron and nickel. These metals are liquid in the outer core ...

... but solid in the inner core! This is because the weight of Earth's other layers puts so much pressure on the inner core.

Sphere 2: Hydrosphere

Welcome to the wettest of the spheres! This includes all the water on and beneath my surface and in the atmosphere.

WATER VAPOR GAS

LIQUID WATER

SOLID ICE

Earth is the only planet in our solar system with a good supply of liquid water on its surface.

Hi! I'm Ocean. Most of Earth's liquid water is salt water found in me.

I'm Fresh Water. I only make up about 3 percent of Earth's water. I'm found in rivers, lakes, glaciers, and under the ground.

All living things need fresh water to drink.

You're welcome!

I'm also used for recreation, cooking, cleaning, sewage systems, and to generate electricity in power plants and hydroelectric dams.

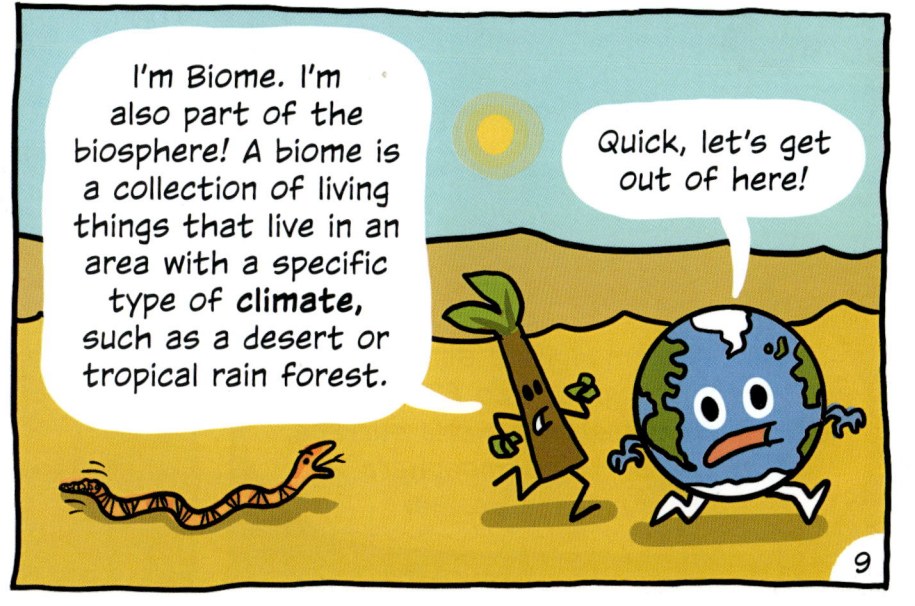

Sphere 3: Biosphere

This includes all living things on the planet...

from the tiniest microorganism...

... to the largest animal that ever lived—the blue whale!

It also includes all plant life...

... and humans, of course!

I'm Biome. I'm also part of the biosphere! A biome is a collection of living things that live in an area with a specific type of **climate**, such as a desert or tropical rain forest.

Quick, let's get out of here!

9

Sphere 4: Atmosphere

And now the fourth sphere...

My turn! I'm Atmosphere. I include all the air that surrounds Earth.

I'm mostly made up of the gases nitrogen and oxygen, with small amounts of argon and other gases, including carbon dioxide, water vapor, methane, hydrogen, and ozone.

NITROGEN (78%)

OXYGEN (21%)

ARGON (AROUND 1%)

OTHER GASES (TRACE AMOUNTS)

Earth's atmosphere stretches from the ground to 62 miles (100 kilometers) above the surface. This end of the atmosphere is where space begins.

There wouldn't be any life on Earth without me! I provide air for living things to breathe, and I protect Earth from the sun's harmful rays.

All the spheres are linked to each other and affect each other.

In the water cycle, water from the hydrosphere evaporates into the atmosphere and then falls back to the surface and moves across the lithosphere and the biosphere.

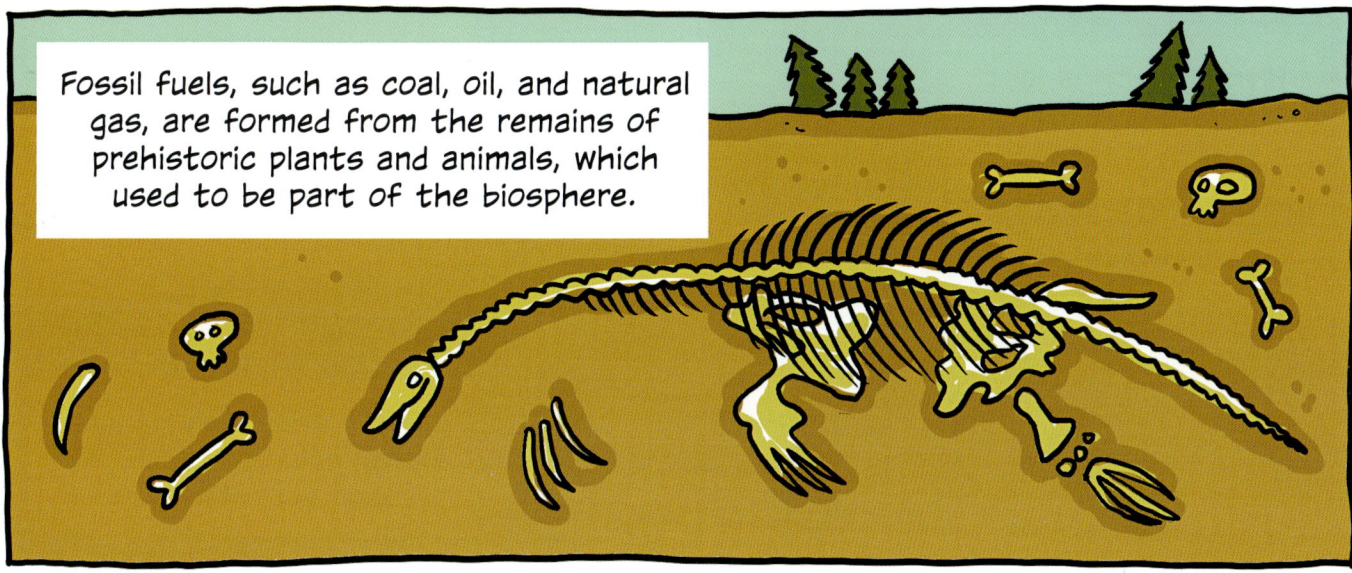

Fossil fuels, such as coal, oil, and natural gas, are formed from the remains of prehistoric plants and animals, which used to be part of the biosphere.

Today, fossil fuels are found in the ground as part of the lithosphere.

The burning of fossil fuels adds carbon dioxide to the atmosphere. This leads to **global warming**, which affects the biosphere and the hydrosphere.

EARTH'S PLACE IN SPACE

From down here on the surface, the planet seems pretty big!

But out here in space, it doesn't seem quite so large!

Planet Earth is just one part of our solar system.

THE SUN
MERCURY
MARS
EARTH
VENUS
JUPITER
SATURN
URANUS
NEPTUNE

There are eight planets here, including Earth, and one star—the sun.

The solar system is itself just a small part of the Milky Way **galaxy** ...

OUR SOLAR SYSTEM

... which is just a tiny part of the universe!

The universe is made up of everything (all matter) that exists anywhere in space and time.

Earth is the third closest planet to the sun. It is about 93 million miles (150 million kilometers) away.

However, light moves so fast that it only takes 8 minutes for the sun's light to travel to my surface!

Heat from the sun makes me warm enough to live on and supplies the energy that plants need to grow.

If we were any closer to the sun, we would be far too hot!

Any further away and we'd be too cold!

Earth and the other planets travel around the sun in **elliptical** paths called **orbits**.

Earth's position in the solar system has an impact on its geography and affects life down here on the surface.

Let's take a closer look ...

13

EARTH'S ORBIT

The sun is the largest item in our solar system by far. It is so big that 1.3 million planets my size could fit inside it (if we could stand the heat)!

The massive amount of matter in the sun gives it a very strong gravitational pull. This force is what keeps Earth and the other planets in orbit around the sun.

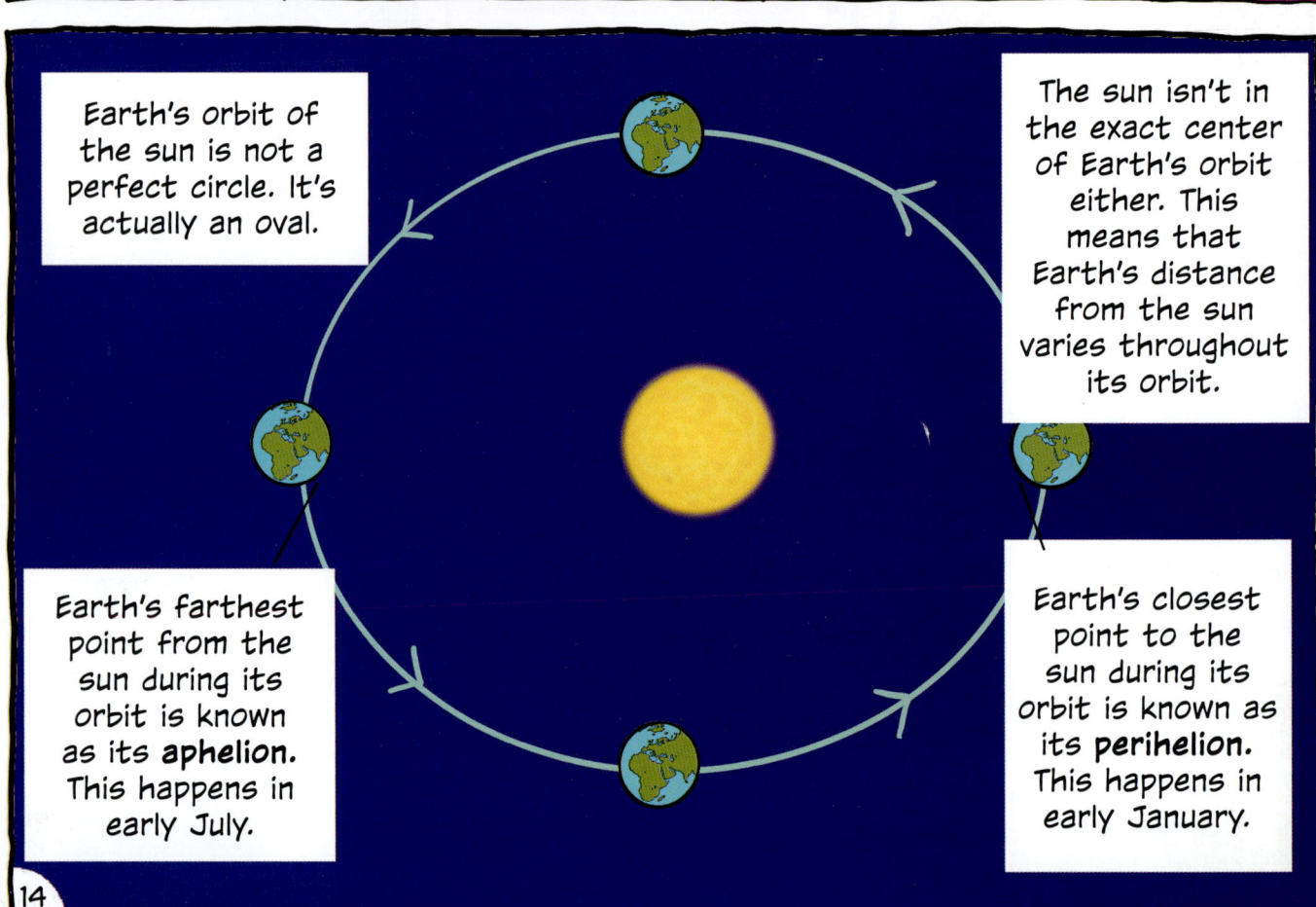

Earth's orbit of the sun is not a perfect circle. It's actually an oval.

The sun isn't in the exact center of Earth's orbit either. This means that Earth's distance from the sun varies throughout its orbit.

Earth's farthest point from the sun during its orbit is known as its **aphelion**. This happens in early July.

Earth's closest point to the sun during its orbit is known as its **perihelion**. This happens in early January.

It takes 365 days, 6 hours, 9 minutes, and 9.54 seconds for the Earth to complete one orbit around the sun.

TICK TICK

However, we consider one year to be 365 days. Over time, all of the extra hours, minutes, and seconds would add up and put our calendar out of sync!

Summer? Winter? I'm confused!

The solution? Adding an extra day at the end of February every four years!

Having 366 days in a leap year keeps our calendar aligned with the Earth's movement.

EARTH'S ROTATION

"As well as moving around the sun, I spin around on my own, just like a top!"

"Luckily we can't feel it, or we'd all be seriously dizzy!"

"Earth spins on its axis, which is an imaginary line that passes through the North and South Poles."

It takes 24 hours for Earth to spin once on its axis. This period of time is known as one Earth day.

DAYTIME

NIGHTTIME

During a day, different parts of the planet experience day and night as they face toward and away from the sun.

The side of Earth facing the sun experiences daytime. It receives light and heat energy from the sun.

The other side of Earth experiences nighttime. The sun's rays can't reach this area because it is facing toward space.

The length of day and night changes throughout the year around the world because Earth is tilted on its axis.

But more on that later! Too early!

Earth's **rotation** also affects me!

It makes air in the atmosphere curve toward the right in the Northern Hemisphere and move toward the left in the Southern Hemisphere.

This is known as the **Coriolis effect**.

The Coriolis effect has an impact on wind patterns on the surface ...

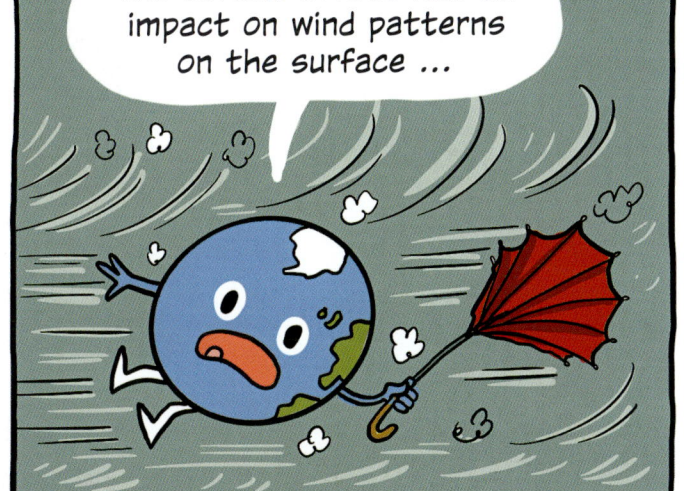

... which in turn creates currents in the ocean!

EARTH'S TILT AND LATITUDE

Did you know that I tilt on my axis? I tilt at an angle of about 23.5 degrees from an imaginary line perpendicular to my orbit!

This means that one **hemisphere** is pointed toward the sun at one point in Earth's orbit. Half a year later, the other hemisphere is angled toward the sun.

This creates seasons. But more on that later!

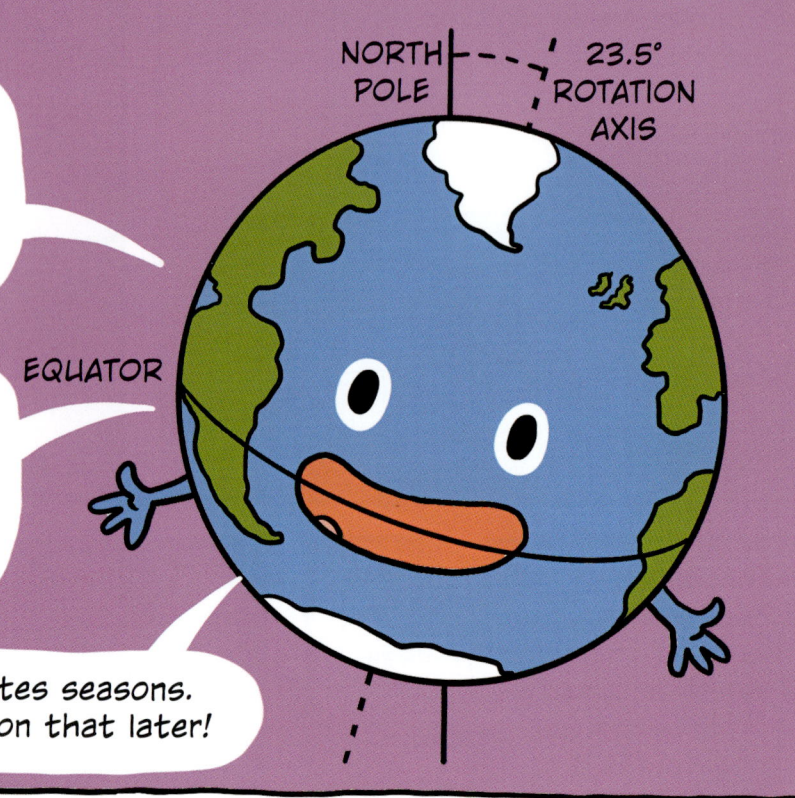

NORTH POLE — 23.5° ROTATION AXIS

EQUATOR

NORTHERN HEMISPHERE

SOUTHERN HEMISPHERE

All land to the north of the equator is known as the Northern Hemisphere. The land south of the equator is called the Southern Hemisphere.

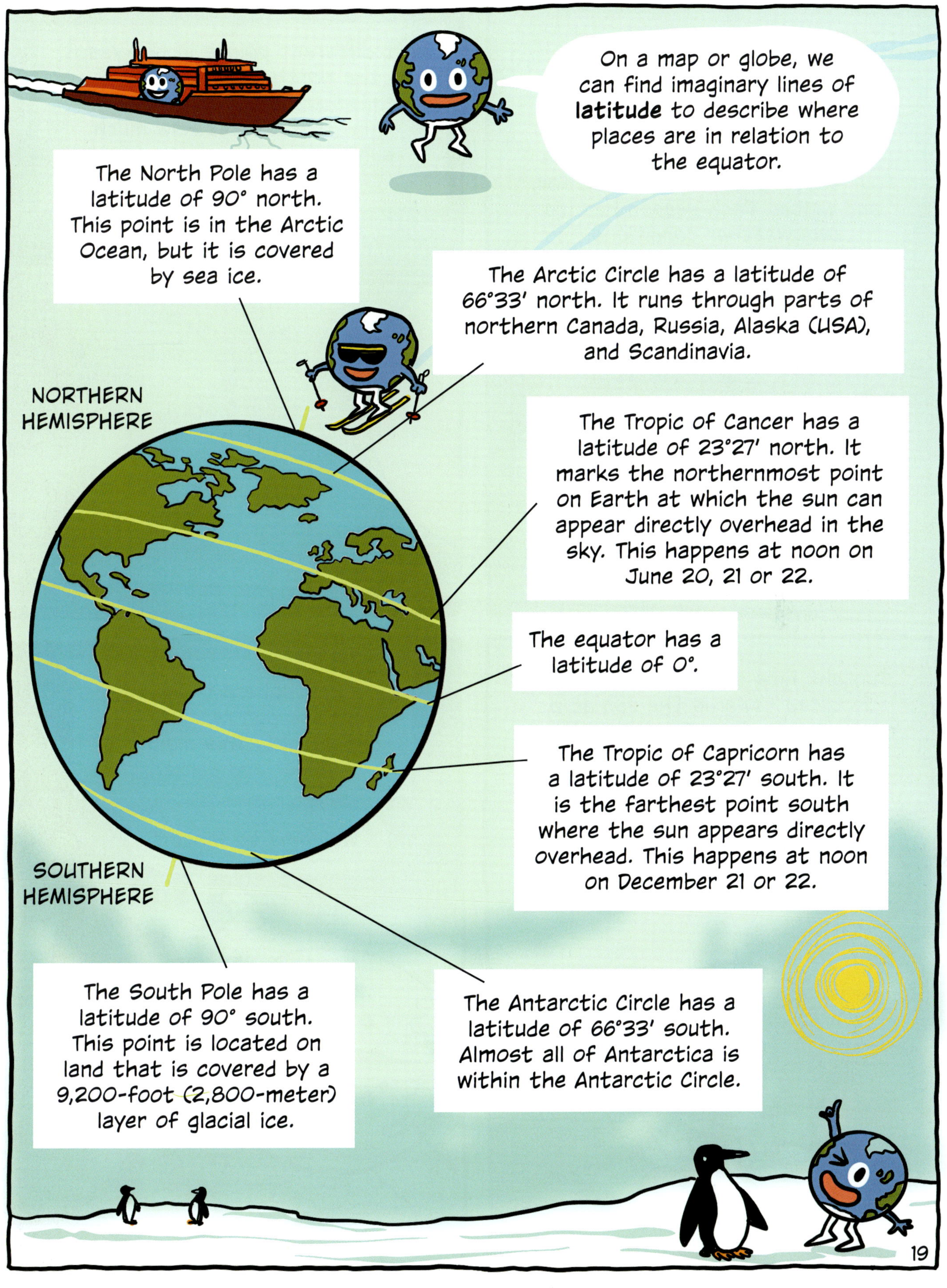

On a map or globe, we can find imaginary lines of **latitude** to describe where places are in relation to the equator.

The North Pole has a latitude of 90° north. This point is in the Arctic Ocean, but it is covered by sea ice.

The Arctic Circle has a latitude of 66°33' north. It runs through parts of northern Canada, Russia, Alaska (USA), and Scandinavia.

NORTHERN HEMISPHERE

The Tropic of Cancer has a latitude of 23°27' north. It marks the northernmost point on Earth at which the sun can appear directly overhead in the sky. This happens at noon on June 20, 21 or 22.

The equator has a latitude of 0°.

The Tropic of Capricorn has a latitude of 23°27' south. It is the farthest point south where the sun appears directly overhead. This happens at noon on December 21 or 22.

SOUTHERN HEMISPHERE

The South Pole has a latitude of 90° south. This point is located on land that is covered by a 9,200-foot (2,800-meter) layer of glacial ice.

The Antarctic Circle has a latitude of 66°33' south. Almost all of Antarctica is within the Antarctic Circle.

19

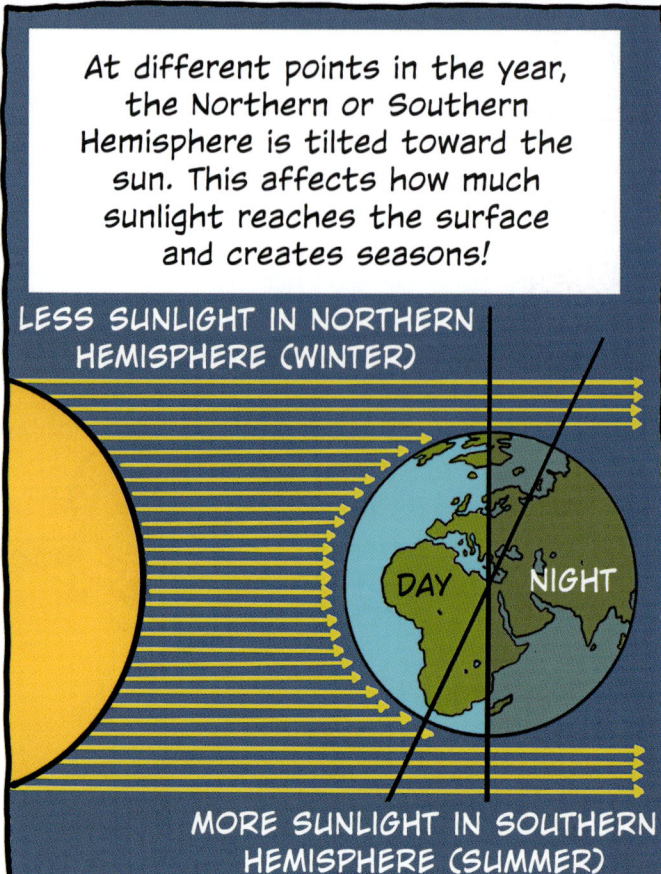

SEASONS

Ready to talk seasons now? Many places on Earth experience four seasons—spring, summer, fall, and winter. Each season has its own weather conditions.

At different points in the year, the Northern or Southern Hemisphere is tilted toward the sun. This affects how much sunlight reaches the surface and creates seasons!

LESS SUNLIGHT IN NORTHERN HEMISPHERE (WINTER)

DAY — NIGHT

MORE SUNLIGHT IN SOUTHERN HEMISPHERE (SUMMER)

Sunlight hits the hemisphere that is tilted towards the sun at a more direct angle. This means that the energy is more concentrated, and so more warmth and light reach the surface.

It's summertime!

The hemisphere tilted away from the sun receives less direct light. This makes it colder and darker.

Welcome to winter!

20

"The hemisphere facing the sun changes as Earth moves through its orbit."

"This is why the seasons change throughout the year!"

JUNE 20, 21 OR 22
The Northern Hemisphere is tilted towards the sun. It is the first day of summer there. The Southern Hemisphere is tilted away from the sun. It is the first day of winter there.

MARCH 19, 20 OR 21
Spring begins in the Northern Hemisphere, and fall begins in the Southern Hemisphere.

DECEMBER 21 OR 22
The Northern Hemisphere is tilted away from the sun, and it is the first day of winter. It's the first day of summer for the Southern Hemisphere, which is now tilted towards the sun.

SEPTEMBER 22 OR 23
This is the first day of fall for the Northern Hemisphere and the first day of spring in the Southern Hemisphere.

In many places on Earth, conditions change with the seasons.

During summer, there are more than 12 hours of sunlight a day. Many people enjoy recreation in the warmer weather. Plants grow well with the long hours of sunlight.

In the fall, temperatures drop and the days are shorter. Some trees lose leaves.

Animals collect seeds, berries, and fruits to store away for winter. Some bird species migrate to warmer places for the winter.

Winter is the coldest season. It's also the darkest, with less than 12 hours of sunlight each day. Trees, except for evergreens, lose their leaves.

Some animals, such as chipmunks, enter a sleeplike state called hibernation to protect themselves from the cold and reduce their need for food.

Even though it's cold and dark, people still find ways to enjoy wintery weather!

In spring, temperatures begin to warm up and the days become lighter.

Migrating birds return to enjoy the warmer weather and increased food supply.

New life begins as new shoots and leaves grow and many animals reproduce.

Welcome back!

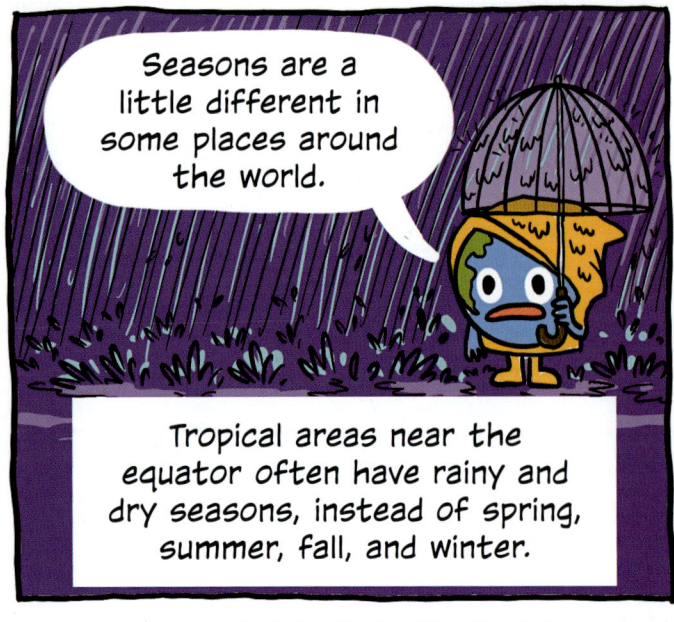

Seasons are a little different in some places around the world.

Tropical areas near the equator often have rainy and dry seasons, instead of spring, summer, fall, and winter.

Climate (long-term pattern of weather) is also different around the world. Many factors affect an area's climate, such as its latitude, as well as the change of seasons—caused by my tilt and my place in my orbit.

The same amount and concentration of sunlight hit the equator all year round. The climate there is always warm, and there are always 12 hours of daylight.

12 hours exactly!

Latitudes that are far away from the equator have much longer days in summer and more hours of darkness in winter. On the Arctic Circle, there are 24 hours of sunlight around mid-June!

Can you believe it's midnight?

And at the poles, it's even more extreme! The sun doesn't rise at all during the winter, and it never sets during the summer!

Hellooooo?

24

There are more things that affect a location's climate. Other factors include distance from the sea ...

... **altitude** (the height of a point in relation to sea level or ground level) ...

... **topography** (the arrangement and heights of the land's surface features) ...

... and wind patterns.

Because of different altitudes, topography, and wind patterns, locations along the same latitude can have different weather conditions! For example, the top of Mount Rainier in Washington, United States, is always covered in snow, even when there is warm summer weather at the base!

EARTH'S MOON

You can't miss my moon! It's the brightest object in the night sky, even though it doesn't produce its own light. It just reflects light from the sun.

The moon is our closest neighbor in space! It is only around 239,000 miles (385,000 kilometers) from Earth. 30 Earth-sized planets could fit in this space.

For something so close, the moon is quite mysterious!

Is it made out of Swiss cheese?

Nope! I found out the hard way that the moon is made of rock, just like Earth.

The moon doesn't have an atmosphere, so there is no wind or weather. There's no protection from **meteoroids** either, so collisions are common. These crashes leave behind **craters**.

The moon's surface is covered with hills, valleys, and mountain ranges, just like Earth!

However, there is no life or liquid water here, so the landscape looks a little different.

Watch out!

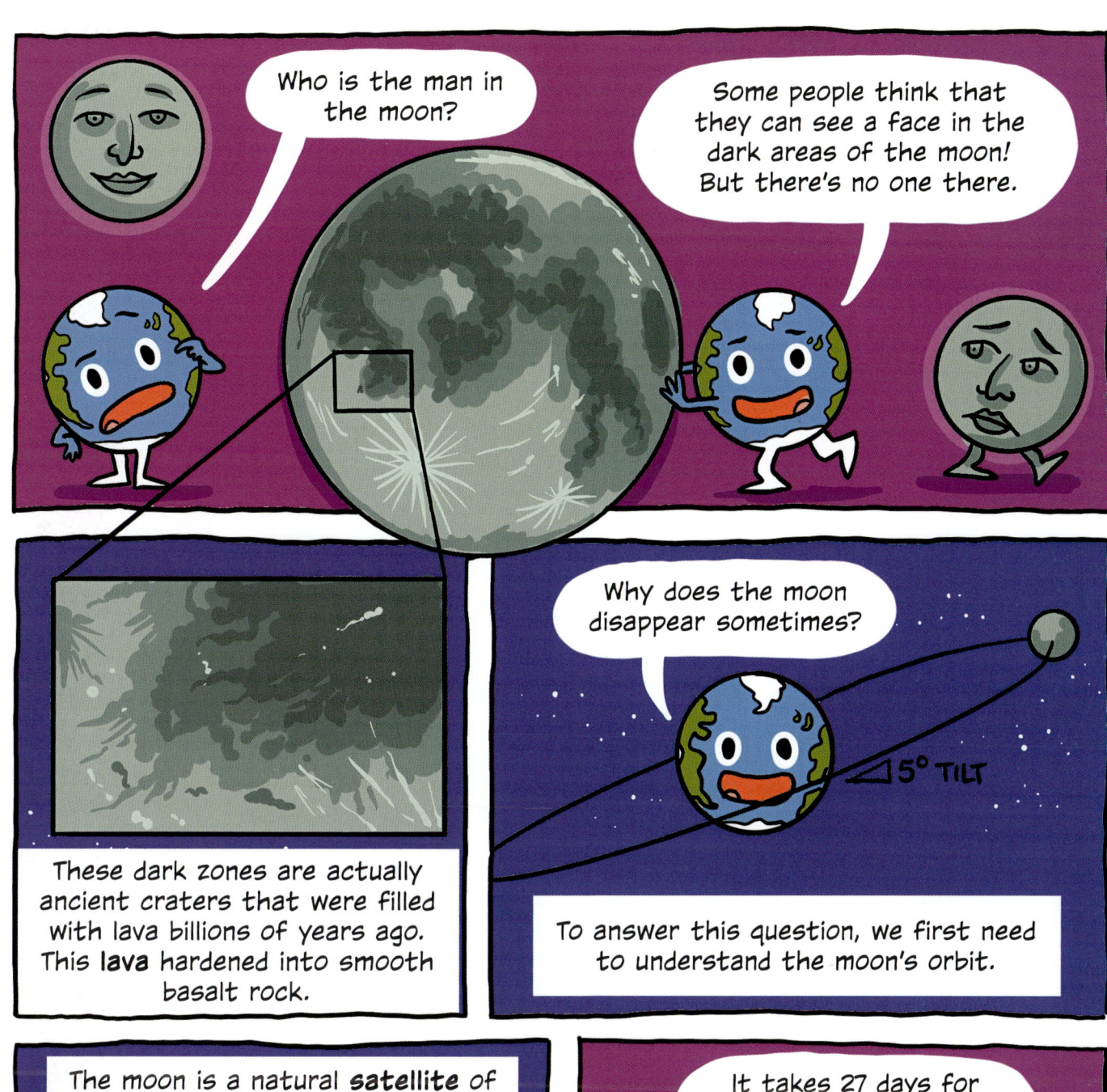

Who is the man in the moon?

Some people think that they can see a face in the dark areas of the moon! But there's no one there.

These dark zones are actually ancient craters that were filled with lava billions of years ago. This **lava** hardened into smooth basalt rock.

Why does the moon disappear sometimes?

5° TILT

To answer this question, we first need to understand the moon's orbit.

The moon is a natural **satellite** of Earth. It is held in orbit by gravity from our planet.

It takes 27 days for the moon to complete one trip around Earth.

JANUARY

However, because Earth is also rotating, the moon's orbit appears to last 29 days instead!

27

Sunlight hits different parts of the moon as it orbits Earth.

The changing pattern of sunlight that hits the moon as it moves through its orbit makes it look as if it's changing shape!

Now here's the answer to the question: The moon seems to disappear when it's between the sun and Earth. This is because, at that time, no sunlight illuminates the side facing Earth.

But it's still there, I promise!

What is the dark side of the moon? Is it always dark there?

We always see the same side of the moon from Earth. This is because the time taken for the moon to rotate on its axis is the same time taken for it to orbit the Earth.

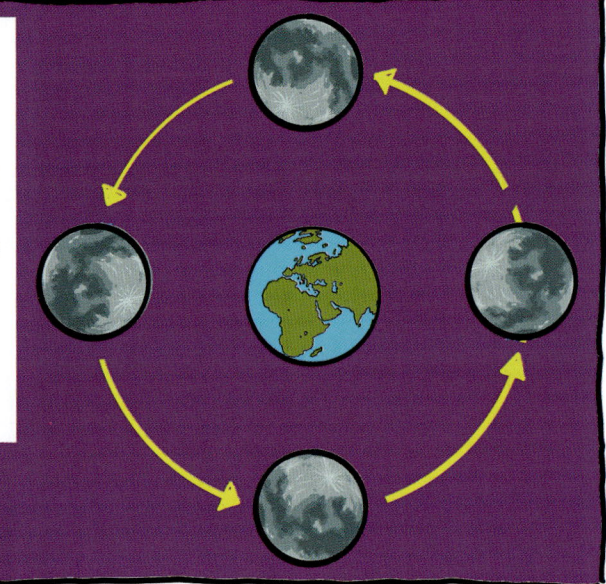

The other side of the moon, which we can't see from Earth, is often called the dark side. However, "far side" is more accurate because both sides of the moon receive sunlight at different times. We just can't see the other side from Earth.

Astronauts have taken photos of the far side of the moon from space.

Nothing mysterious there at all ... wink, wink!

28

What's fascinating is that the moon also affects the geography on Earth! It reflects light from the sun, giving moonlight at night. Without it, nights would be very dark. But that's not all.

The moon's gravity helps to hold Earth in place. Without that, the tilt of the planet might wobble and tilt too far or hardly tilt at all. This could result in no seasons or very extreme seasons and wild weather. Days would be shorter.

Stay ... still ...!

The combined force of gravity from the moon and the sun pulls on the oceans to cause high tides on the sides closest and furthest from the moon, and low tides in the areas between. Without the moon, there would be no tide changes.

Some animals' behavior is linked to the lunar cycle of the moon orbiting Earth. Once a year, right after a full moon, all the corals on the Great Barrier Reef reproduce on the same night!

Barau's petrel birds use the moon to **synchronize** their migration. The birds always arrive at their breeding grounds together on the night of a full moon.

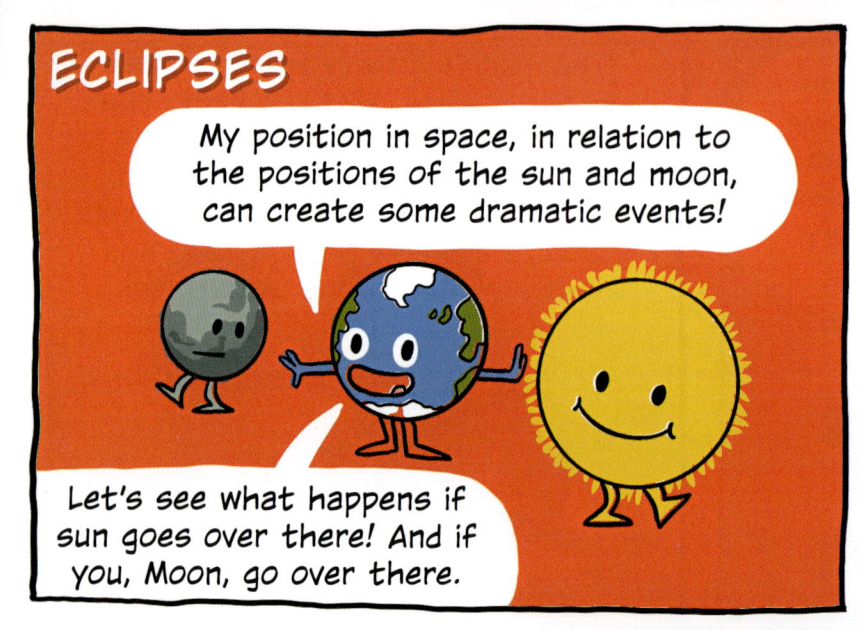

ECLIPSES

"My position in space, in relation to the positions of the sun and moon, can create some dramatic events!"

"Let's see what happens if sun goes over there! And if you, Moon, go over there."

"Wow! This is a total **lunar eclipse**. It's sometimes known as a blood moon because the moon appears to turn red!"

During a lunar eclipse, Earth moves into a position where it blocks the sun's light so that it can't hit the moon.

SUN — EARTH — MOON

However, Earth's atmosphere still bends some sunlight towards the moon. The atmosphere scatters the other colors of light into space, leaving just red light to hit the moon!

SUN LIGHT — BLUE LIGHT — RED LIGHT

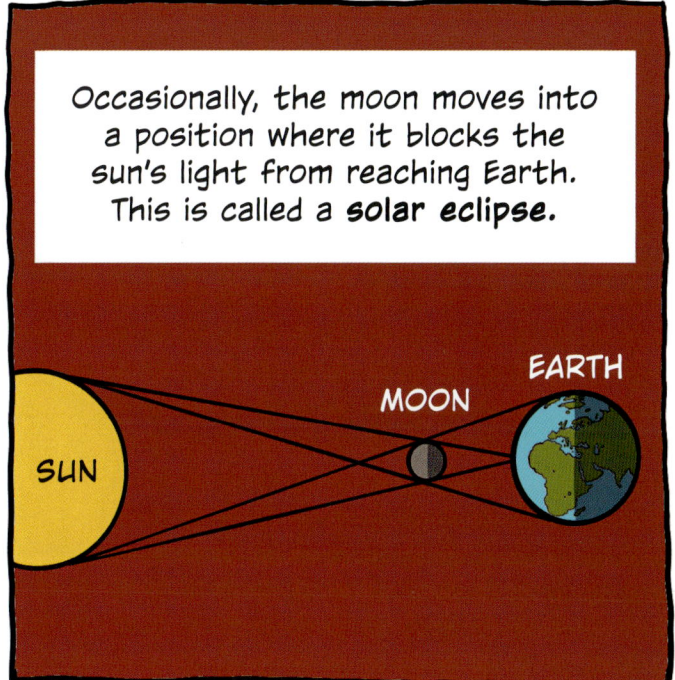

Wait, where's the sun going now? I can't keep up!

Occasionally, the moon moves into a position where it blocks the sun's light from reaching Earth. This is called a **solar eclipse.**

SUN MOON EARTH

During a total solar eclipse, the moon covers the whole of the sun, leaving just a bright ring around the edges. Everything goes dark for a few minutes, even though it's the middle of the day!

You should use glasses with a special eclipse filter or a pinhole projector to look at a solar eclipse. Even though most of the sun's rays are blocked, the visible rays can permanently damage your eyes.

HUMANS IN SPACE

Since the beginning of human history, we have gazed up at the stars with wonder and curiosity.

People, including scientists, began to form theories about Earth's place in space and how the universe worked.

Some of these ideas weren't quite right.

Many people used to think that Earth was at the center of the universe and the sun and other planets orbited around it!

Eventually, humans developed telescopes to observe space without ever leaving the ground.

As early as 1610, the Italian astronomer Galileo Galilei was able to spot four of Jupiter's moons through his telescope!

Our understanding of space changed forever when astronauts traveled into space for the first time in the 1960's. They took the first photo of Earth from space ...

... saw the far side of the moon for the first time ...

... and even set foot on the moon's surface!

Since then, robotic spacecraft have traveled through our solar system and beyond.

Humans have sent **space probes** to other planets, asteroids and comets.

Earthlings have even used rovers to explore the surface of Mars!

33

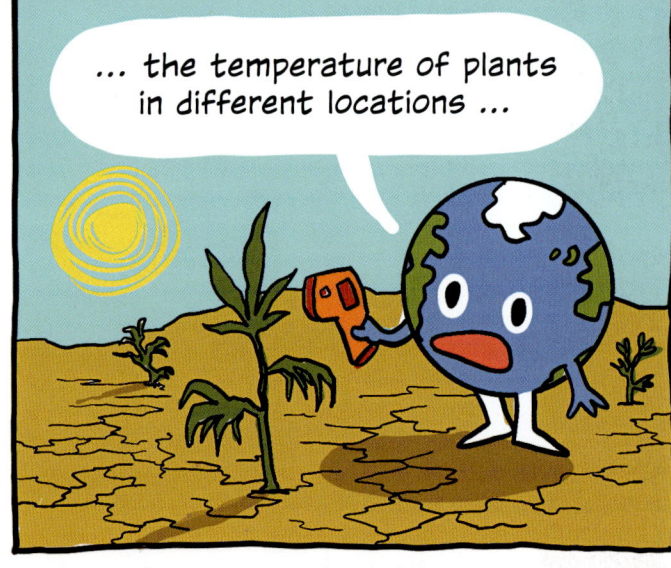

By looking at the planet from space, everyone can learn a lot about life on Earth. How great it is that many satellites in orbit around me monitor and track geography down on the surface!

These satellites contain **sensors** that can tell us how much carbon dioxide is in the atmosphere ...

... the temperature of plants in different locations ...

... and which areas have the most air pollution.

The satellites are also in the perfect spot to study weather conditions, such as lightning, and to track storms.

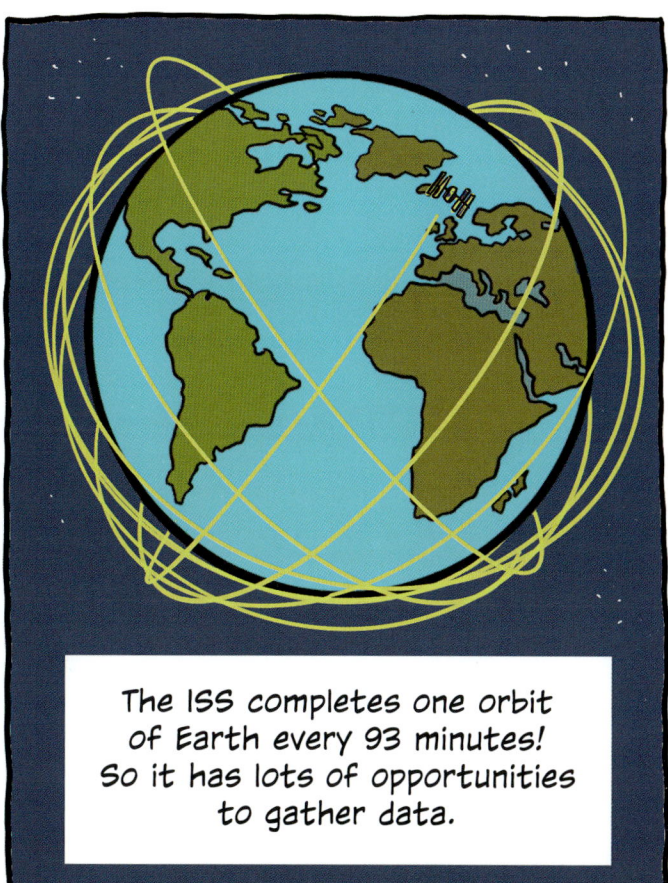

There are many sensors on the International Space Station (ISS). This is a large satellite where astronauts can live.

The ISS completes one orbit of Earth every 93 minutes! So it has lots of opportunities to gather data.

Seeing Earth from up here reminds me how precious and unique I am!

We haven't yet found anywhere else in the universe where humans could live. So we have to protect Earth and its natural environment so that our home is safe for many, many years to come!

ACTIVITY: EXPLORE THE PHASES OF THE MOON!

Sunlight hits different parts of the moon as it orbits Earth. The changing pattern of sunlight that hits the moon as it moves through its orbit makes it look as if it's changing shape!

These different shapes are called *phases*.

Why does the moon have phases? See for yourself by doing this experiment.

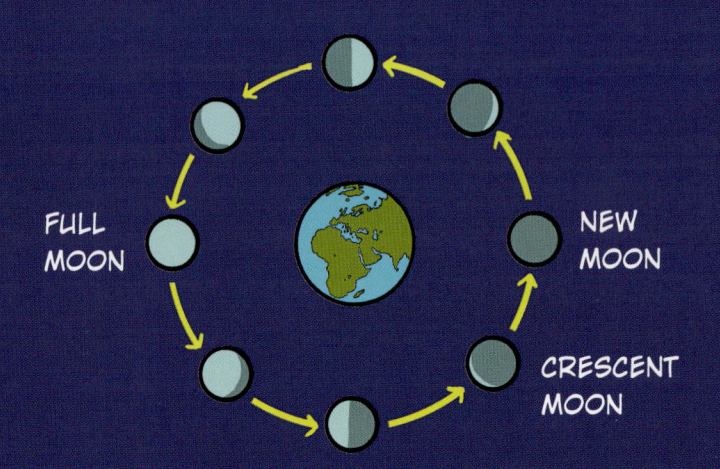

What You'll Need
- A flashlight
- A small ball (a tennis ball or baseball will do)

1. Place the flashlight on a table or shelf. Set the ball on a surface of the same height. Then shine the light on it. The flashlight acts as the sun. The ball is the moon. You are the planet Earth.

Sit directly between the light and the front of the ball, but beneath the beam of light. The whole side of the ball facing you will be in the light like a full moon is.

FULL MOON

36

2. Move a little to the "side" of the ball. You will see half of the ball in the light, like a half moon.

HALF MOON

3. Move around the ball a little more, so that the ball is nearly between you and the light. Most of the ball will be in shadow. Only a small part will be in the light, like a crescent moon.

CRESCENT MOON

Now you know why the moon has different phases. The moon and Earth change positions in relation to the sun.

One phase is often called the "new moon." This is when the moon does not appear to be visible in the night sky. Think about why that is. Where can you stand in relation to the flashlight and ball to simulate the new moon phase?

Answer: If you stand behind the ball and the flashlight, the ball will block out the beam from the flashlight. When the moon is between Earth and the sun, the moon is invisible in the night sky. We call this the new moon!

Find a calendar that shows the dates when the different phases of the moon appear. Then you can observe the phases of the moon yourself in the night sky!

CAN YOU BELIEVE IT?!

Earth rotates at about 1,000 miles per hour (1,609 kilometers per hour)!

Earth's rotation is slowing down! The rotation is slowing by about 17 milliseconds per hundred years. That means that in about 140 million years the length of a day will increase to 25 hours!

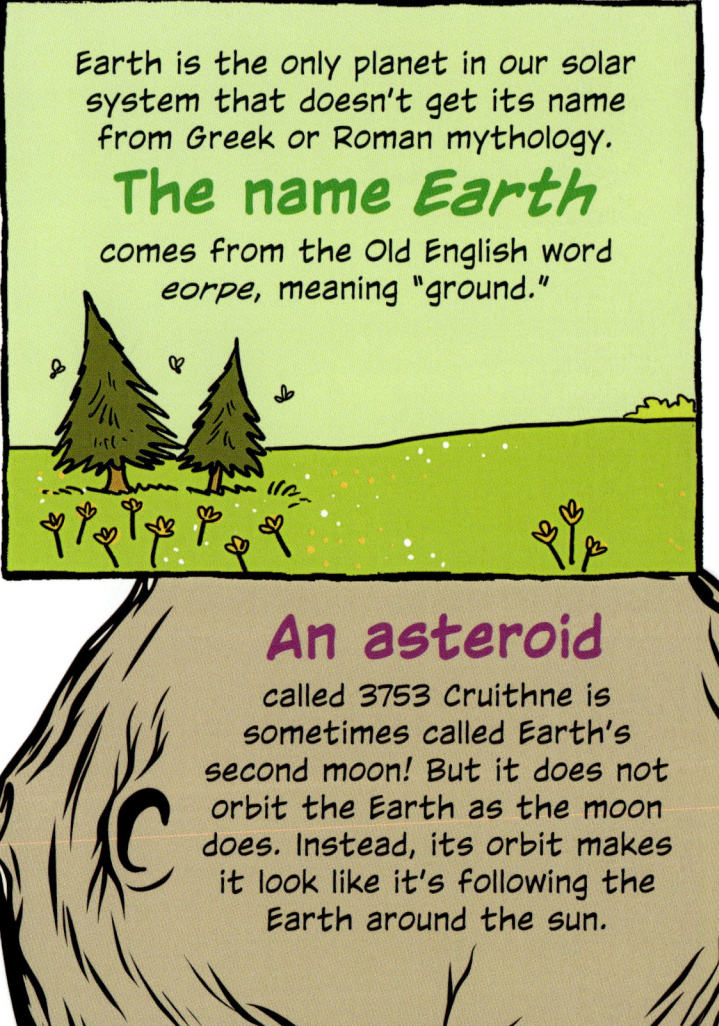

Plants lose their leaves in autumn due to the **decreasing length of the daylight,** not the cooler temperatures of the season.

Earth is the only planet in our solar system that doesn't get its name from Greek or Roman mythology. **The name Earth** comes from the Old English word *eorpe*, meaning "ground."

An asteroid called 3753 Cruithne is sometimes called Earth's second moon! But it does not orbit the Earth as the moon does. Instead, its orbit makes it look like it's following the Earth around the sun.

A prehistoric map

of the night sky and stars in Lascaux Cave in France shows that people thought about Earth's place in space more than 16,500 years ago!

Every year, there are at least

two solar eclipses

visible from Earth. They occur in different locations.

Earth moves around the sun

at 67,000 miles per hour (107,826 kilometers per hour)!

WORDS TO KNOW

altitude the height of a point in relation to sea level or ground level.

aphelion the position of a planet when it is furthest from the sun.

axis an imaginary line through the center of an object that is spinning.

circumference the distance around a circular object or sphere.

climate the weather patterns of an area in general or over a long period of time.

Coriolis effect the deflection of circulating air caused by Earth's rotation. This affects the motion of the winds and ocean currents.

crater a deep hole in the ground.

crust the outermost shell of Earth's surface.

elliptical shaped like an oval.

galaxy a collection of stars, planets, gas and dust held together by gravity.

global warming the increase in the average temperature on Earth.

gravity a force that attracts objects towards each other and pulls objects towards the ground on Earth.

hemisphere one of the two halves of Earth.

latitude the position north or south of the equator.

lava molten (melted) rock that pours out of a volcano or crack in the earth.

lunar eclipse when the moon becomes darker or red because Earth's shadow moves across it.

meteoroid a piece of rock from space.

microorganism a living thing that is too small to be seen without a microscope.

orbit the curved path of an object in space around a planet or star.

perihelion the position of a planet when it is closest to the sun.

rotation movement in a circle around a fixed point.

satellite a natural or artificial object in space that revolves around a planet.

sensor a device that gathers data.

solar eclipse when the sun is blocked from view by the moon.

solar system a group of planets that move around a star.

space probe a small spacecraft without any humans on board that collects data in space to send back to scientists.

sphere shaped like a ball.

synchronise to cause to happen at the same time.

topography the physical appearance of the natural features of an area of land.

universe everything (all matter) that exists anywhere in space and time.